Worm

Jill Bailey

Heinemann Library
Des Plaines, Illinois

© 1998 Reed Educational & Professional Publishing

Published by Heinemann Library,

an imprint of Reed Educational & Professional Publishing,

1350 East Touhy Avenue, Suite 240 West

Des Plaines, IL 60018

Designed by Celia Floyd
Illustrations by Alan Male
Printed in Hong Kong / China

02 01 00 99 98
10 9 8 7 6 5 4 3 2 1
Library of Congress Cataloging-in-Publication Data

Bailey, Jill.
Worm / Jill Bailey.
p. cm. -- (Bug books)
Includes bibliographical references and index.
Summary: A simple introduction to the physical characteristics, diet, life cycle, predators, habitat, and lifespan of worms.
ISBN 1-57572-665-3 (lib. bdg.)
1. Worms--Juvenile literature. [1. Worms.] I. Title. II. Series.
QL386.6.B35   1998
592'.3--dc21   98-10623
CIP
AC

Acknowledgements

The Publishers would like to thank the following for permission to reproduce photographs: Ardea London: J.P. Ferrero p. 5, P. Morris p. 4; Bruce Coleman: F. Labhardt p. 10, Dr. F. Sauer p .5, K. Taylor p. 22; FLPA: G. Hyde pp. 14, 20, M. Rose p. 12, M. Thomas p. 25; Garden Matters: K. Gibson p. 24, J. Phipps p. 27; Chris Honeywell p. 28; NHPA: D. Woodfall p. 7; Oxford Scientific Film: K. Atkinson pp. 21, 29, J. Cooke pp. 15, 26, B. Davidson p. 11, C. Milkins p. 13, R .Redfern p. 23, H. Taylor p.19, D. Thompson pp. 8, 9, 16, 17, 18

Cover photographs reproduced with permission of Chris Honeywell

Every effort has been made to contact copyright holders of any material reproduced in this book. Any omissions will be rectified in subsequent printings if notice is given to the Publisher.

Any words appearing in the text in bold, **like this**, are explained in the Glossary.

Contents

What are worms?

Worms are long, thin, and soft. Their head is rounded and their tail is more pointed. Worms do not have legs.

A **saddle** around the worm's middle makes a slippery slime called **mucus**. This helps worms slip easily through soil. There are different kinds of worm. We are going to look at earthworms.

Where do worms live?

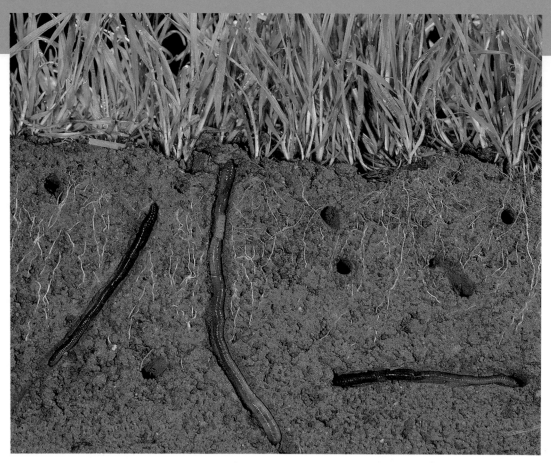

Worms live in **burrows** in soil. Usually they live near the surface. In very dry or cold weather they may tunnel deeper than the height of a grown-up.

A pile of soil from your backyard and as high as you may contain twenty worms. A pile of soil from an old field may contain over 700 worms!

What do worms look like?

A worm has no eyes
or ears. Its whole body
can taste and it can feel light and
sound. You can see the long, thin tube
that carries red blood through the worm.

A worm's body is made up of many **segments.** Each segment has a few small, stiff **bristles** that act like tiny hooks. They grip the sides of the **burrow** to pull the worm along.

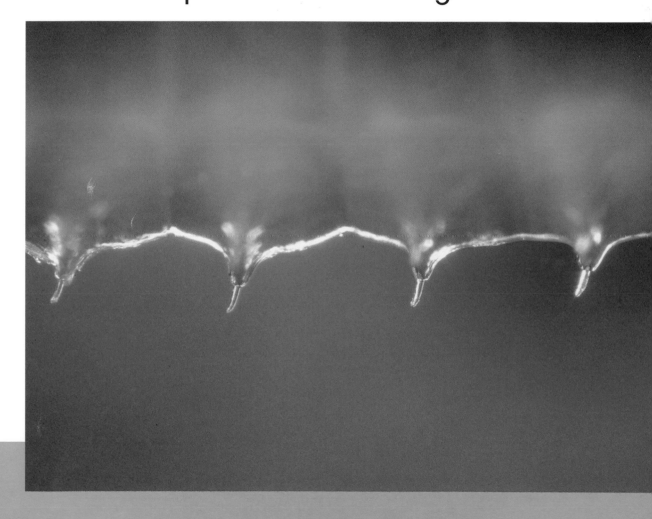

How big are worms?

A small earthworm can be as long as your finger. A long earthworm can stretch from your fingertips to your elbow. The smallest earthworm is only as long as your smallest fingernail.

The biggest earthworms in the world can be the length of four grown-ups lying end to end. They can be as wide as two fingers side by side.

What do worms do?

Worms spend most of their time underground. They push their way through soil, or dig with their mouths.

On warm, damp nights worms may come to the surface of the soil. They drag dead leaves back into their **burrows** to eat. Have you seen leaves sticking out of worm burrows?

How long do worms live?

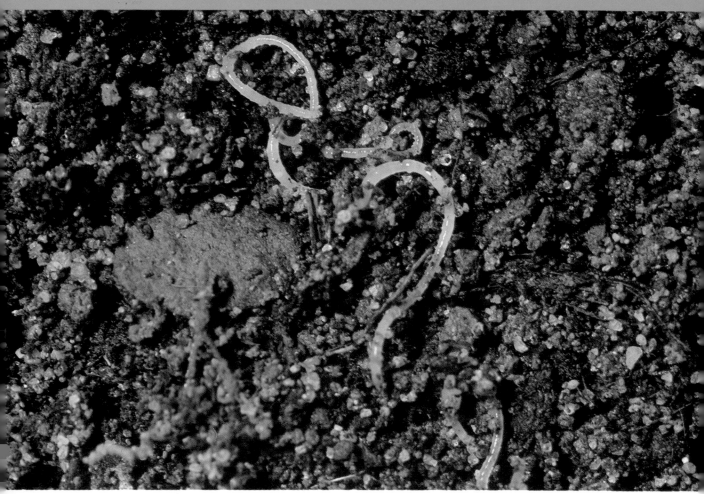

A baby worm takes about a year to grow big enough to have its own babies. It may then live for another ten years.

Worms cling so hard to their **burrows**
that if birds try to pull them out, their
bodies may break in half. Sometimes
one half grows back into a worm.

How are worms born?

On warm, damp summer evenings, worms come out of their **burrows** and lie side-by-side to **mate**. Each worm gives its partner a drop of special juice. This will help eggs to grow.

Each worm makes a thick **mucus** with its **saddle** to lay its eggs in. The mucus turns hard to make a **cocoon** for the eggs. You can see a cocoon here.

How do worms grow?

A worm may lay up to twenty eggs inside each **cocoon**, but usually only one survives. The cocoon keeps it safe until it is ready to **hatch**.

After one to five months the baby hatches from the cocoon. It is tiny and stays hidden in the soil. It will take at least a year to grow as big as its parents.

What do worms eat?

Worms eat parts of dead plants and animals they find in the soil. They also eat the soil as they tunnel.

They **grind** down the soil to get at its food. The useless grains of soil pass out of the worm's tail end. Sometimes the waste soil makes small piles.

Which animals attack worms?

Many birds, mice, badgers, and other animals are the worm's predators. They listen for the sound of the worm moving, then they dig it up.

There are also predators underground. Moles dig tunnels through soil to look for worms. Moles cannot see in the dark but they can hear and smell.

How are worms special?

Worms are important to farmers and gardeners. They break down dead leaves by eating them. This puts **minerals** back into the soil which help plants grow.

Each worm can move about 25 times its own weight of soil a year. Worm **burrows** let air and rain into the soil. This keeps the soil healthy.

How do worms move?

The worm's soft body is filled with a watery liquid. The worm can squeeze its body into different shapes. To move forward, the worm makes its front end long and thin.

Then it digs its **bristles** into the ground and pulls its tail end forward. Now it digs in the bristles on its tail and lets go of the front ones as it stretches forward again. And so on.

Thinking about Worms

See for yourself how a worm changes its shape. Fill a balloon almost full of water and tie it up. Now squeeze it. It will get long and thin.

The worm is pulling leaves into its **burrow**. What will it use them for? How has it made its burrow? How does the worm get rid of the soil it digs out?

Worm Map

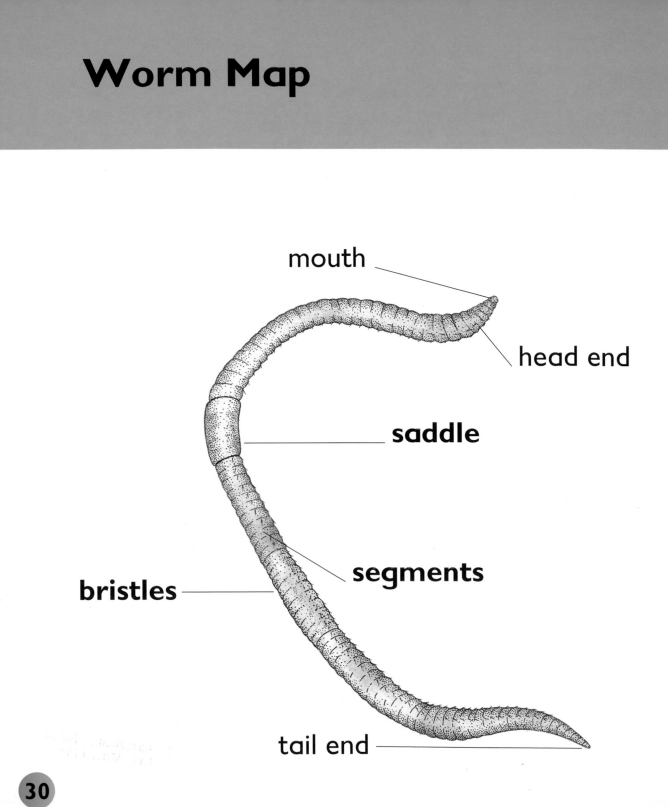

mouth

head end

saddle

segments

bristles

tail end

Glossary

bristles small, very stiff hairs that help the worm to grip the soil as it moves

burrow a hole that an animal makes in the ground

cocoon a tiny hard case that protects the eggs and baby worms until they hatch

grind to crush into smaller pieces

hatch to be born out of an egg or **cocoon**

mate when two animals come together to make babies

minerals special food that animals and plants need to live

mucus slimy, sticky stuff that helps a worm to slip through the soil and also makes the **cocoon**

saddle a slimy band around the middle of the worm. It makes the **mucus** that protects the worm's eggs.

segment parts of a worm that are alike

More Books to Read

Kalman, Bobbie & Schaub, Janine. *Squirmy Wormy Composters*. New York: Crabtree Publishing. 1992

Van Laan, Nancy. *The Big Fat Worm*. New York: Random House. 1995

Index